# YOUR KNOWLEDGE HAS VALUE

# The irrelevance of the location of Riemann's zeros to the disposition of the prime numbers

William Fidler

**Bibliographic information published by the German National Library:**

The German National Library lists this publication in the National Bibliography; detailed bibliographic data are available on the Internet at http://dnb.dnb.de.

ISBN: 9783346634153
This book is also available as an ebook.

© GRIN Publishing GmbH
Nymphenburger Straße 86
80636 München

Print and binding: Books on Demand GmbH, Norderstedt, Germany
Printed on acid-free paper from responsible sources.

GRIN web shop: https://www.grin.com/document/1191150

## The irrelevance of the Riemann zeros to the disposition of the prime numbers

W M Fidler

# List of Contents

# Abstract

It is shown here that there is a direct connection between the Riemann zeros, the counting numbers, and hence the prime numbers, but not in the so-called Critical Strip. We regard the Critical Strip as an ugly thing, in which there is a singularity (**at x** = 1), none of the positions of the zeros, **Im(s)**, on the line of symmetry can be determined with absolute precision, and it is not even certain that all of the zeros lie on the line of symmetry [1]. We have described elsewhere [2] that we regard the Critical Strip as a mathematical graveyard, and the work performed here merely reinforces that conclusion. It is difficult to reconcile the statement 'the closer that the real part of the zeros lies to ½, the more regular the distribution of the primes'[3] with what has been determined in this work, for it shows that there is no causal link between the Riemann zeros and the counting numbers, but rather, the opposite.

A mathematical structure is developed in which the articulation of the numerical location of a Riemann zero in the sequence of zeros is sufficient to determine the counting number with which it is associated, its location, **Im(s)**, on the line of symmetry and, if scaled by $\pi$, the Gauss/Legendre prime number counting function associated therewith.

All of this is made possible by examining a 'critical strip' with a line of symmetry passing through any one of the oft-dismissed trivial zeros. It should be borne in mind that Euler's formula relating the zeta function to the sum over the primes was not conditional upon the nature of the field in which it was derived, and it was only upon Riemann's extension through analytic continuation that the behaviour of the zeta function throughout the whole of the complex plane could be investigated.

The location of a zero of the zeta function on any of these 'critical lines' can be determined with absolute precision for it only requires the articulation of a counting number. This is in contradistinction to the effort required to determine even approximately, only one of the locations of a zero on the Critical Line.

We show that the positions of Riemann zeros on anything other than a critical line in any of the strips, critical or otherwise is of no consequence in determining the magnitude of the prime number counting function.

It is concluded that the disposition of the prime numbers within the system of counting numbers is solely an intrinsic characteristic of that system and is totally unrelated to the distribution of the Riemann zeros.

# Introduction

Riemann's 1859 paper [4], was concerned with the determining of the number of prime numbers in any given range of numbers. In the course of the work he noticed that in the so-called Critical Strip all of the zeros of the zeta function should lie on the line of symmetry, $x = 1/2$. He glossed over attempting any serious resolution of this observation for he considered it incidental to the object of the paper.

Since then some of the best minds in mathematics have, without success, attempted to verify Riemann's conjecture and many subsequent theorems in number theory are based upon its veracity.

The investigations of the number and location of the zeros in the Critical Strip have involved much and complex mathematics and, indeed, the fixation of concentration on the Critical Strip to the exclusion of any other regions in the complex plane. The identification of zeros in the negative half of the complex is an easy task, and, such zeros have been christened trivial and of little importance. However, it has been shown recently [2], at least on the ordinates passing through these trivial zeros that there are an infinite number of zeros at well-defined positions; indeed, it is shown that there is a zero at infinity on all of these lines.

It has been suggested by the author [5] and others that the Riemann conjecture may be undecidable and, indeed, it was considered to be incorrect by J E Littlewood. Hardy and Littlewood [6] did show that there are an infinite number of zeros on the Critical Line, although it is shown in [5] that there is not a zero at infinity, but rather, the negative of the Dirichlet alternating function for $x = \frac{1}{2}$ .

# Analysis

Whilst much of the following has appeared elsewhere by the author [2] its repetition here is considered essential to the arguments presented later in the work.

The Riemann zeta function, $\zeta(s)$ is an extension to the series:

$$\zeta(s) = \frac{1}{1^s} + \frac{1}{2^s} + \frac{1}{3^s} + \frac{1}{4^s} + \frac{1}{5^s} + \;- - - - - \;------------\; (1).$$

Here, the real number exponent is replaced with a complex number, $s = x + i\,y$.

It should be noted that we use Riemann's notation for the complex number but the normal mathematical notation for its real and imaginary parts.

Under the same constraint as above we write the Dirichlet eta function, $\eta(s)$ as :

$$\eta(s) = \frac{1}{1^s} - \frac{1}{2^s} + \frac{1}{3^s} - \frac{1}{4^s} + \frac{1}{5^s} \;- \;------------------\; (2).$$

From equations (1) and (2) we get:

$$\zeta(s) - \eta(s) = 2^{1-s}\left[\frac{1}{1^s} + \frac{1}{2^s} + \frac{1}{3^s} + \frac{1}{4^s} + \;- - - -\right] = 2^{1-s}\,\zeta(s).$$

For reasons which will become apparent later in the analysis we write the above as:

$$-\eta(s) = (2^{1-s} - 1)\,\zeta(s) \;------------------------------(3).$$

Now, Riemann's functional equation is: $\zeta(s) = 2^s \pi^{s-1} \sin\left(\pi\,{}^s/_2\right) \Gamma(1-s)\zeta(s-1)$.

It then follows that we may write the Dirichlet functional equation in terms of the Riemann functional equation

Hence, $\eta(s) = (1 - 2^{1-s})2^s \pi^{s-1} \sin\left(\pi\,{}^s/_2\right) \Gamma(1-s)\zeta(s-1)$. ---------------- (4)

We now seek a solution to equation (1) when $\zeta(s) = 0$.

Equation (1) is written out in extenso :

$$\zeta(s) = \frac{1}{1^{x+iy}} + \frac{1}{2^{x+iy}} + \frac{1}{3^{x+iy}} + \frac{1}{4^{x+iy}} + \frac{1}{5^{x+iy}} + \;------------------\; (5).$$

For the sake of illustration consider the second term of the above.

i.e. $\quad \frac{1}{2^{x+iy}} = \frac{e^{-iy\,\ln 2}}{2^x}$ , which by Euler's theorem may be written:

$$\frac{1}{2^{x+iy}} = \frac{1}{2^x}\left[\cos(y\,\ln 2) - i\,\sin(y\,\ln 2)\right].$$

Hence, we may collect terms in equation (5) and write:

$$\zeta(s) = \sum_{k=1}^{k=\infty} \frac{1}{k^x} \cos(y \, lnk) - i \sum_{k=1}^{k=\infty} \frac{1}{k^x} \sin(y \, lnk) \text{ ----------------------------- (6).}$$

If we set $y \, lnk = k \, \pi$ it then follows that all of the imaginary terms disappear. This leaves the real part to be given by: $\sum_{k=1}^{k=\infty} \frac{1}{k^x} \cos(k \, \pi)$.

Hence, the real part of equation (6) becomes : $- \frac{1}{1^x} + \frac{1}{2^x} - \frac{1}{3^x} + \frac{1}{4^x} - \frac{1}{5^x}$ ------

This may be written: $- [\frac{1}{1^x} - \frac{1}{2^x} + \frac{1}{3^x} - \frac{1}{4^x} + \frac{1}{5^x}$ ------].

But, this is equal to $-\eta(x)$.

If the real part of (6) is to disappear then from equation (3) we have:

$$-\eta(x) = (2^{1-x} - 1) \, \zeta(x) = 0.$$

Either the first term on the RHS disappears or the second.

The first term will vanish if $x = 1$, but this would reduce equation (1) to the Harmonic series which is known to diverge, albeit slowly, to infinity. The product above would then be of the form: $0 \cdot \infty$, which is indeterminate. It then follows that we must take $\zeta(x)$ to be zero.

We now return to the functional forms for $\zeta$ and $\eta$.

Again, Riemann's functional equation is: $\zeta(s) = 2^s \pi^{s-1} \sin\left(\pi \, s/2\right) \Gamma(1 - s) \zeta(s - 1)$.

Now, the sine term will vanish if we set $s = -2n$, where $n$ is a real integer.

It then follows, that setting $s = -2n$ will yield the following result: $\zeta(-2n) = 0$, and hence, from equation (3), $\eta(-2n)$ will also vanish. It is important to emphasize that the sine term will vanish if, and only if, n is an integer.

This procedure outlined is, of course, that employed in generating the so-called 'trivial zeros' of the Riemann zeta function. Further, from the functional equation we see that whatever $\zeta$ is evaluated to on the left of $s = \frac{1}{2}$ is determined by its evaluation on the same point reflected across $s = \frac{1}{2}$. This, together with the analytic continuation of $\eta$ provides the ability to compute $\zeta$ anywhere in the complex plane. In addition, since $\zeta(s)$ has no zeros to the right of **Re(s) = 1**, then the functional equation predicts that there are no other non-trivial zeros to the left of **Re(s) = 0** and hence all of the non-trivial zeros lie within the Critical Strip.

The arguments set out as follows are the same for all 'critical strips' centered on the trivial zeros; let us consider a 'critical strip' centered on the first trivial zero at x = -2.

We have shown [2] that, on the 'critical line' a Riemann zero is located at every ordinate point, y, given by the equation $y = k\pi/\ln k$. Since k = 1,2,3,4,----, and we have shown [5] that there exists horizontal lines which we have named Dirichlet lines and along which Riemann's zeta function $R(s)$ is given by, $R(s) = -\eta(x)$, we may then state that, at the ordinate given by the above formula there is a Riemann zero associated with each of the natural numbers; it then follows that there must be a Riemann zero associated with each prime number.

### The assigning of values of the prime number counting function to the Riemann zeros along the line passing through x = ½

From [7] we obtained the magnitude of the imaginary part of the coordinate of the first Riemann zero and which is truncated to **14.134725**. In the following illustration it is again shown how a value of the prime number counting function may be linked to a zero of Riemann's zeta function

Let it be assumed that the magnitude of the imaginary part of s may be calculated from the simple formula: $y = k\pi/lnk$ , where y = 14.134725 = Im(s).

We rearrange the above equation in iterative form, i.e. $k_{n+1} = (y/\pi)lnk_n$ ---------------- (7).

Hence for the example here, $k_{n+1} = 14.134725/\pi. lnk_n = \Omega \, lnk_n$ --------------------(8).

Using an initial value of 10, equation (8) iterates rapidly to yield **k = 10.637895**, although, all of the digits in the decimal part are superfluous for our purpose.

It follows that the point in question is 'bracketed' by the Dirichlet lines for **k = 10** and **k = 11** which intersect the so-called Critical Line at **s = x = ½** at $R(s) = -\eta(1/2)$ as shown in Fig1

**Note**: It is found that all subsequent iterative calculations shown herein may be started with an initial value of**10.**.

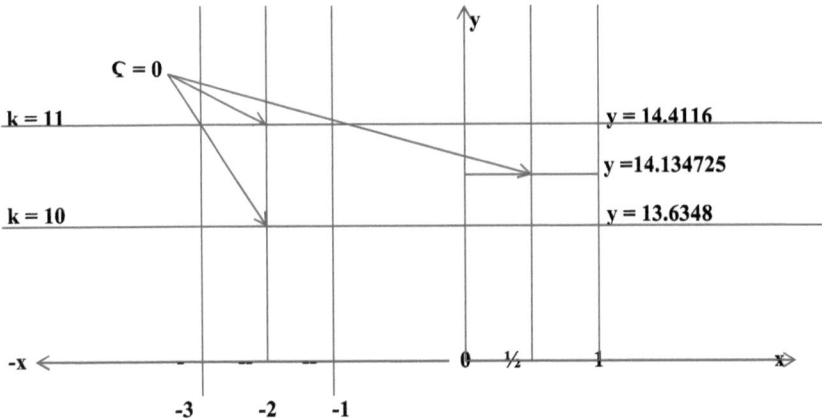

**Fig1**

Now, the integers, **k = 10** and **k = 11** each have an associated value of the prime number counting function (indeed, they may be the same). Further, we discard the upper value of **k** for it makes no sense to evaluate the number of primes associated with the point of interest because the location of the Riemann zero there is less than that for the zeta function at **k = 11**. In addition, the lower bound is an even number and there are, with the exception of **2** no even prime numbers. Hence, for the purpose of associating a value of the prime number counting function we take the lower nearest odd number (which may, or may not, be prime), and which, in this case is **9 (y = 12.8682)**. For brevity we will, in all further illustrations refer to a number of this nature as a **nearodd** and denoted by the symbol, **d**. Thus, every 'Riemann zero' and its associated value of the prime number counting function will have a corresponding nearodd which is located within the matrix, Table1, developed in [8] (where the primes are presented in bold and underlined), or, from Kulsha's data [9].

| | | | | | | | | | | | | | | |
|---|---|---|---|---|---|---|---|---|---|---|---|---|---|---|
| **3** | **5** | **7** | 9 | **11** | **13** | 15 | **17** | **19** | 21 | **23** | 25 | 27 | **29** | **31** |
| 33 | 35 | **37** | 39 | **41** | **43** | 45 | **47** | 49 | 51 | **53** | 55 | 57 | **59** | **61** |
| 63 | 65 | **67** | 69 | **71** | **73** | 75 | 77 | **79** | 81 | **83** | 85 | 87 | **89** | 91 |
| 93 | 95 | **97** | 99 | **101** | **103** | 105 | **107** | **109** | 111 | **113** | 115 | 117 | 119 | 121 |
| 123 | 125 | **127** | 129 | **131** | 133 | 135 | **137** | **139** | 141 | 143 | 145 | 147 | **149** | **151** |
| 153 | 155 | **157** | 159 | 161 | **163** | 165 | **167** | 169 | 171 | **173** | 175 | 177 | **179** | **181** |
| 183 | 185 | 187 | 189 | **191** | **193** | 195 | **197** | **199** | 201 | 203 | 205 | 207 | 209 | **211** |
| 213 | 215 | 217 | 219 | 221 | **223** | 225 | **227** | **229** | 231 | **233** | 235 | 237 | **239** | **241** |
| 243 | 245 | 247 | 249 | **251** | 253 | 255 | **257** | 259 | 261 | **263** | 265 | 267 | **269** | **271** |
| 273 | 275 | **277** | 279 | **281** | **283** | 285 | 287 | 289 | 291 | **293** | 295 | 297 | 299 | 301 |
| | | | | | | | | | | | | | | |
| | | | | | | | | | | | | | | |
| 1 | 2 | 3 | 4 | 5 | 6 | 7 | 8 | 9 | 10 | 11 | 12 | 13 | 14 | 15 |

**Table 1**

We then find from the data of A V Kulsha [9] that the value of the prime number counting function associated with the nearodd of the first Riemann zero is equal to **4,** and this is verified by inspection of the matrix, which only contains the odd numbers and so does not take account of the first prime, **2**. Note that the lowest row of numbers denotes the column numbers of the matrix.

It noted that if the sum of the digits of the nearodd is divisible by **3** then we may move to the next nearodd;  e.g. if it is calculated for some **y** that equation (8) converges to a value of 99.2361789 (numbers simply chosen at random by the author) then strictly, the nearodd is 99. But, the sum of the digits is divisible by **3** and so we proceed to the next nearodd, which is 97, and from Kulsha's data [9] we see that there are 25 primes including that point. This is verified by counting the primes in the matrix above.

# The impossibility of an ordinate of any Riemann zero in the Critical Strip to be directly associated with an integer

The proof of this is simple, for if the iterative equation (8) was to yield an integer for any of the values of the ordinate of a Riemann zero in the Critical Strip (published, or otherwise) then this would mean that the ordinate lay on a Dirichlet line, and whilst this would render the imaginary part of equation (6) equal to zero, the real part, $Re[R(s)]$ would equal $-\eta(x)$, which is only zero at the intersection of a Dirichlet line with an ordinate line passing through any of the trivial zeros. Further, it follows that, in the context of the prime number counting function, this means that all of the Riemann zeros in the Critical Strip must lie between two Dirichlet lines, and the numbers associated with them, as determined from equation (8) cannot contribute to the magnitude of the prime number counting function for they are not integers. To illustrate this consider Table2 and Fig2. Table2 consists of the first 10 Riemann zeros (i.e. the ordinates lying on the line through $x = \frac{1}{2}$, and taken, suitably truncated, from [7] ) with the 'raw' results, $k_{n+1}$ as determined from equation (8), the nearodd $d$, and the magnitude of the prime number counting function $R(d)$, taken from [9]. Fig2 shows part of Table2 laid out on the line, s = ½.

| Im(s) | $\Omega$ | $k_{n+1}$ | d | R(d) |
|---|---|---|---|---|
| 14.1347 | 4.4992 | 10.6380 | 9 | 4 |
| 21.022 | 6.6915 | 20.068 | 19 | 8 |
| 25.018 | 7.9612 | 25.911 | 25 | 9 |
| 30.4249 | 9.6845 | 34.211 | 33 | 11 |
| 32.9351 | 10.4836 | 38.186 | 37 | 12 |
| 37.5861 | 11.9640 | 45.737 | 45 | 14 |
| 40.9187 | 13.0248 | 51.283 | 51 | 15 |
| 43.3271 | 13.7914 | 55.355 | 55 | 16 |
| 48.0052 | 15.2805 | 63.408 | 63 | 18 |
| 49.77383 | 15.8435 | 66.498 | 65 | 18 |

### Table 2

**Note.**

We could have economized the work in the above table by recognizing that the sum of the digits of some of the nearodds is divisible by **3**, but we refrained from so doing in the interest of clarity.Fig2, which is not to scale, and uses only some of the data presented in Table2 clarifies the arguments presented previously, viz: that none of the numbers associated with Riemann's zeros in the Critical Strip is an integer.

$$s = 1/2$$

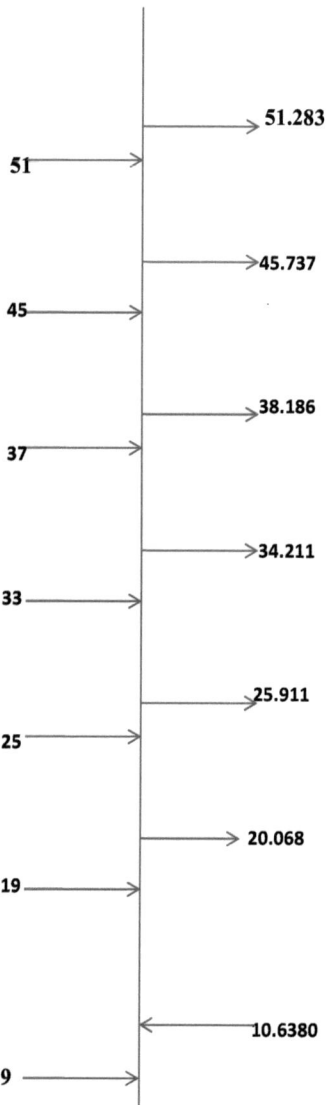

**Fig2**

# A critique of the importance attributed to the Critical Strip

Consider an abstract of the graph of the prime number counting functions against the nearodds, as shown in Fig3.

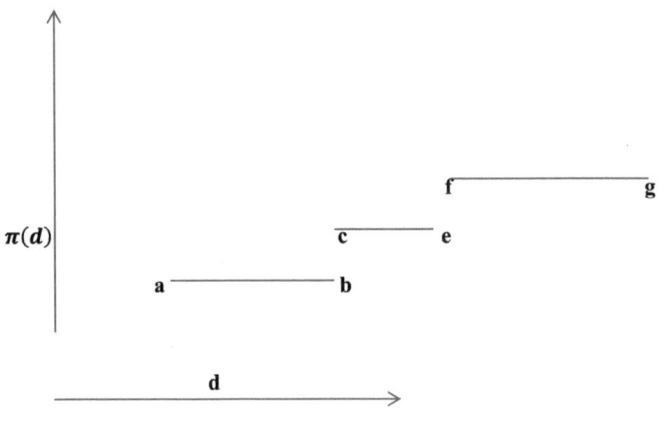

## Fig3

The graph of the prime number counting function must only consist of a series of horizontal lines spaced apart in the vertical direction by one unit. In the diagram **a** and **b** are taken to be prime numbers, as are **c** and **e**, and **f** and **g**, with **b** = **c** and **e** = **f**.

There can be no connection of any kind between **b** and **c** for example, for non-integer values of the prime number function have no meaning, and, if **a** and **b** are the only prime numbers on the line **ab**, then any other numbers, even, odd, and non-integer in the interval between **a** and **b** contribute nothing to the magnitude of the prime number counting function.

It is in the spirit of the foregoing that we offer the following criticism of the concentration of attention over the last 160 years on Riemann's Critical Strip.

We have shown here that none of the Riemann zeros in the Critical Strip can be associated with an integer; in addition the herculean efforts involved in the determination of **Im(s)** show that it is not even certain exactly where the Riemann zeros are located. This is exemplified by perusing Odlyzko's data [7] where **Im(s)** is calculated to one thousand places of decimals—it is not stated that the calculations terminated at this stage or that further computation would not make any appreciable difference to **Im(s)**.

Hence, in the Critical Strip it appears that none of the **Im(s)** can be determined with a precision that would put beyond doubt that the exact position of the Riemann zero had been established, and, none of the Riemann zeros can be directly associated with an integer--- hence the reason for the devising of the idea of a nearodd. It is not certain that all of the Riemann zeros lie on the Critical Line; it was determined by Conrey [1] that **40%** of the zeros lie on the line and this amount has recently been superseded—although not by much.

## Discussion and conclusion

In the light of the statements and analysis made here it is concluded that, whilst the Riemann zeros in the Critical Strip are interesting, in that it seems that they all line on the line passing through $x = \frac{1}{2}$ they do not contribute to the prime number counting function, for such a function is only concerned with integers and we have shown here that none of the Riemann zeros on the Critical Line can be directly associated with an integer. Whilst it is legitimate to ask, for example,' what is the magnitude of the prime number counting function associated with the number 98.764523182.........', the question is answered by asking 'what is the magnitude of the prime number counting function associated with the number 97'.

Each Riemann zero on an ordinate passing through any one of the trivial zeros has an integer associated directly with it. Any zeros lying within the cells that are formed by two adjacent Dirichlet lines contribute nothing to the magnitude of the prime number counting function. Since there are an infinite number of Riemann zeros along these ordinate lines through any of the trivial zeros then there must be an infinite number of zeros directly associated with all of the counting numbers, in particular, the odd numbers; there is, of course, a zero associated with the number **2**, the first of the primes and the only one which is an even number.

The location, **y** of these zeros on the line is given by the formula $y = k\pi/\ln k$, and it hence follows that y is a function of k, and not vice versa (although the equation may be rearranged as in equation (8)); in addition, it has been noted previously by the author that the above formula shows that the position of any Riemann zero on an ordinate passing through a trivial zero is directly proportional to the Gauss/Legendre prime number counting function. We then see that, along these ordinate lines the position of the Riemann zero only depends upon an integer, and the zero is not responsible for the position of that number in the hierarchy of numbers, and so, the fact that the prime numbers are disposed as they are is not connected with the Riemann zeros, for their disposition is solely a characteristic of the system of numbers. It follows that, in the critical strips in the negative half of the complex plane the Riemann zeros on the line of symmetry can be distinguished from each other, for each is directly associated with a unique counting number.

We then conclude that the disposition of the Riemann zeros which are associated with the prime numbers and lie within the infinite set of Riemann zeros associated with all of the counting numbers is exactly the same as the disposition of the prime numbers within the natural numbers. Indeed, if we denote the position of any Riemann zero within the sequence of the zeros on a line passing through any of the trivial zeros by a number, then that identifier is also the integer with which the zero is associated.

It then follows that Table1, which is a part of the matrix containing all of the odd numbers is also a matrix of all of the Riemann zeros associated with the odd numbers, and, of greater import if we identify the position of a zero in the sequence of zeros, then we may locate this precisely within the matrix

14

If, for example we wished to determine the location within the matrix of the odd number associated with the **203rd** zero in the sequence of zeros then it is plain that this number is also **203**; we have shown previously, [10], how to determine the number of prime numbers and their location in any row of the matrix and so we may determine the exact location of the zeros associated with these primes. Hence, from Table1, the zeros numbers which are associated with the prime numbers in this row are **191**, **193**, **197**, **199** and **211**.

We can form four matrices of the same size as that shown in Table1, where the odd number 'leaf' is overlaid with the Riemann zeros laid out in the same sequence as the sequence of the odd numbers, followed by the location **Im(s)** on the ordinate through a trivial zero and finally by scaling **Im(s)** by $\pi$, recover the Gauss/Legendre prime number counting function. Projecting downwards from any zero will give the corresponding odd number, then upwards to yield the value of **Im(s)**, then upwards again to give the Gauss/Legendre prime number counting function. All of this is shown in Fig4.

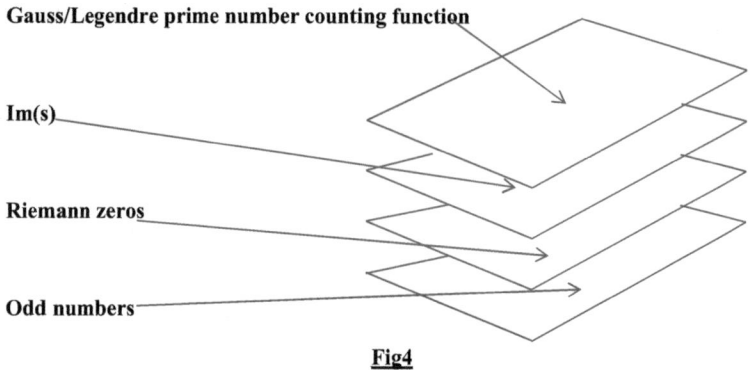

**Gauss/Legendre prime number counting function**

**Im(s)**

**Riemann zeros**

**Odd numbers**

<u>**Fig4**</u>

We could add another 'leaf' to the above diagram which would incorporate Riemann's prime number counting function given, for our purposes, by the reduced Gram series.

Hence, for the example given above we would have, in addition to, $y_{203} = Im(s) = \frac{203\pi}{\ln(203)}$, and $\pi(203) \cong \frac{203}{\ln(203)}$, $\pi(203) \cong \sum_{n=1}^{n=\infty} \left[\ln(203)\right]^{n} \Big/ \left[nn! \, \zeta(n+1)\right]$.

It follows from all of the foregoing that there are only two 'types' of critical strip in the complex plane; in the negative half there are an infinite number of identical strips, in which it

15

is shown that, along any critical line each counting number (and hence each prime number) is directly associated with a Riemann zero, the position of which may be determined exactly and where we can calculate both the Gauss/Legendre and Riemann prime number counting functions, whilst in the positive half there is only one unique critical strip, where none of the Riemann zeros may be directly associated with a counting number, the exact position of none of the zeros on the critical line can be determined and large computational effort must be expended to ensure that the location has been determined with sufficient accuracy to ensure that a Riemann zero lies in the vicinity of the calculated position. This region is also littered with the corpses of the many failed attempts to validate Riemann's conjecture.

Finally, we conclude that, in the investigation of the relationship between the Riemann zeros, the prime numbers and the prime number counting function, we should focus our attention on any of the ordinates passing through a trivial zero of the Riemann zeta function for it is here that we have established a direct link between the prime numbers and Riemann's zeros.

**W M Fidler**

**March 2022.**

16

# References

[1]    More than two fifths of the zeros of the Riemann zeta function are on the
       Critical Line.

                      J B Conrey

              J. reine angew. Math 399 (1989), 1-26.

[2]    On the Riemann Hypothesis

                      W M Fidler

           GRIN Verlag, ISBN 9783346388575  (2021).

[3]    Riemann's 1859 Manuscript

           Clay Mathematics Institute.

[4]    On the number of primes less than a given magnitude

                    G F B Riemann.

       The Monthly Notices of the Berlin Academy, Nov 1859.

[5]    The assigning of values of the prime number counting function to Riemann's
       Zeros. The concept of the Dirichlet line in the complex plane.

                      W M Fidler

           GRIN Verlag ISBN 9783346583758 (2022).

[6]    The zeros of Riemann's zeta function on the Critical Line

                 GH Hardy & J E Littlewood.

             Math.Z, 10(3-4): 283-317 (1921).

[7]    Tables of the zeros of Riemann's zeta function

                      A Odlyzko

       www.dtc.umn.edu/~odlyzko/zeta-tables/index.html

[8]     Determining the primality of a number by the use of an accelerated version of trial division.

W M Fidler

GRIN Verlag ISBN 9783346493002 (2012).

[9]     The fluctuations of the prime number counting function $\pi(x)$.

Compiled by A V Kulsha

www.primefan.ru/stuff/primes/table.html

[10]     On the disposition of the prime numbers

W M Fidler

GRIN Verlag ISBN9783346548696 (2021).